THE ULTIMATE BEEKEEPER'S LOGBOOK:

A beekeeper's journal to keep track of over 250 hive inspections, supplies, expenses, income, and bee products. Includes recipes, advice, and more! Made especially for new beekeepers and small apiaries producing extra income for the hobbyist beekeeper.

By: T.M. Copeland

Copyright 2022

4th Edition

Dedicated to my Bee Mentor: Jeff F.

Thanks for getting me started!

THE ULTIMATE BEEKEEPER'S LOGBOOK
TABLE OF CONTENTS

THE JOY OF BEEKEEPING	6
TOP TEN QUESTIONS ASKED BY NEW BEEKEEPERS	8
THE LANGUAGE OF BEEKEEPING	11
BEEKEEPING SUPPLIES	13
A YEAR OF BLOOMS	16
APIARY SETUP	19
BEE LIFE CYCLE	22
HOW TO FIND THE QUEEN	30
WHEN YOU CAN'T FIND THE QUEEN	33
COMMON PESTS AND PARASITES OF HONEYBEES	34
A YEAR OF BEEKEEPING	38
INSPECTION LOG	47
HARVEST LOG	95
BEE BUSINESS INCOME	98
SWARM TRAP LOG	104
QUEEN REARING LOG	116
BEE FEEDING RECIPES	124
BEE PRODUCT RECIPES	127
NOTES	136

THE JOY OF BEEKEEPING

If you are reading this you either are interested in becoming a beekeeper, or you have already taken the great leap into the adventure that is beekeeping! To me, beekeeping is a perfect blend of science, nature, community, and sustainability. When I started beekeeping, like most new beekeepers, I was just interested in making honey! I had no idea about all the other products bees personally provide, or anything about all the value-added products I could make from my bee made harvests! All this, and we can't forget all that bees do for our natural environment and agriculture!

I also had no idea that beekeeping would help me get to know my community and neighbors through bee mentoring, selling and bartering, and even by helping to start a beekeeping club for local high school students. As a science teacher and nature lover, I was simply amazed at the biology of bees including their life cycle, behavior, and their will to survive. My first year I had two hives, harvested 40 pounds of honey, 20 cut comb squares and I made creamed honey, candles, lip balm, ornaments, and lotion bars. I sold to neighbors, online, and at a holiday craft market. One of my hives lost their queen, another hive was attacked by mites, and eventually perished in late fall. My daughter helped me with all aspects of beekeeping and we even made several beekeeping videos on our family YouTube channel – Whistle Thicket Farmstead. I started paying attention to what trees and flowers bloomed in my area and when. Plants I had viewed as roadside weeds, suddenly became nectar factories for my bees! After my first beekeeping season, I was hooked!

Unfortunately, there was a problem I noticed after my first season as a new beekeeper, and that was staying organized in my bee yard and keeping track of everything! Although I LOVE beekeeping, there is a lot to learn and a lot to keep track of, especially if you want to turn your hobby into a small income producing side business!

.

There were so many questions to ask and answers to find! What supplies did I need to order for next year, and where could I get the best deal? What flowers were blooming in my area and when did they bloom? What exactly did I need to look for when I did a bee inspection? Did I make a profit selling all my bee made goods? Did I at least break even this year?

The Ultimate Beekeeper's Logbook was the answer to the necessity of keeping track of all this information. This is not just a how to book on beekeeping, but rather an actual logbook for the beginning beekeeper, all the way to hobbyist beekeepers who keep more than a few hives but want to up their game and become more successful and start making an income, or at least become more organized and have all their beekeeping data in one place. When you go to your bee yard, bring this logbook with you to keep track of your data! I hope this logbook helps you on your journey and inspires you to pass on your beekeeping knowledge to others!

- Tom Copeland

Co-Owner of Whistle Thicket Farmstead

TOP TEN QUESTIONS ASKED BY NEW BEEKEEPERS

As a beginning beekeeper, I had so many questions, and so much to learn! After talking with new and aspiring beekeepers, I have compiled the top ten questions most new beekeepers ask and tried my best to answer them!

Question 1: Is beekeeping expensive?

Answer: Well, that depends! The average price for a new hive, protective gear, tools, bees, and a few other items is right around 500.00 dollars. Yes, that is expensive! But, if you have ever raised other farm animals, this price is much less than raising some other types of livestock. Luckily there are some ways to save, including building some of your own hives, borrowing equipment from an established mentor, finding used equipment that is free of damage and disease, and letting family know what to get you for your birthday and holiday gifts! The price also drops once you have several hives, since you only need to buy safety equipment and tools once!

Question 2: Will I get stung?

Answer: Yes, yes you will! Eventually every beekeeper gets stung, and when it happens, you probably will be stung several times! I went almost 6 months before getting my first sting! Just remember, when the sting starts to hurt a bit, that the bee that stung you just lost its life defending a perceived threat to its hive!

Question 3: How much safety equipment should I wear?

Answer: Whatever makes you feel comfortable! New beekeepers tend to wear full protection, whereas other more experienced keepers tend to wear less gear. I have found that I always wear my veil (unless I'm doing a quick peek from the top in the winter), and I always have long pants and a long shirt that is tucked in. When I harvest honey, I find the bees are most angry, and can you blame them? Usually when you are harvesting, the bees are entering a nectar dearth and you are stealing their winter food (at least they see it that way). When I harvest I usually have my bee jacket and bee gloves.

Question 4: Will I get honey my first year?

Answer: Don't plan on it. Most hives can't produce enough extra honey until the second season. The first year is the hardest for a new colony, they must build a lot of drawn comb, which can be reused the next year. I have harvested a decent amount of honey from a first year hive! Every hive is different!

Question 5: What is the difference between a bee package and a nuc?

Answer: A bee package is typically 3 pounds of bees, which is about 10,000 bees, and a mated queen. The queen comes in a queen cage with several worker bees to help feed her. The rest of the bees must chew through a sugar tunnel to release the queen. This takes about 3-5 days, during which time, the 10,000 or so worker bees become accustomed to the pheromones of the new queen. The disadvantage of a bee package is that the queen must start laying, a process which takes about a month until you get new adult bees (see the section on bee life cycles for more details). During this time, some worker bees will die of natural causes, and the population of bees will decline. The queen is also an unproven queen, and although rare, sometimes the worker bees may not accept the queen. The advantages of a bee package are that they are significantly cheaper than a nuc, and much more available throughout the country.

A nuc is a fully functioning hive with all age ranges of bees, a queen that is laying, and larva. This is a mini hive, typically 4-5 frames, which already have drawn comb. The advantage is you already have a colony that is ready to forage on the first day! Because of this, you can get a nuc later in the season and still have a successful hive. The disadvantage is the significant cost difference between a bee package and a nuc, with a nuc being anywhere from 40 to 60 percent more expensive. Nucs are generally less readily available in most areas than packages.

Question 5: Where can I get bees?

Answer: Most bee farms are in southern states, where bees can be raised all year, and winters are mild. The time range for getting bees is the middle of March to the middle of May, depending on your location. If you live in the south, you may be able to drive to the actual bee farm to purchase your bees. Most folks however will be purchasing from large bee farms and meeting in a predetermined location and day to pickup bees. With the growing popularity of beekeeping, most people can find bees in their area. Some companies will even deliver the bees to your post office, maybe even your door. There may even be local beekeepers selling their bee splits, or you may be able to catch a swarm, which are naturalized escaped "wild bees". I would not suggest catching a swarm as your main way to start beekeeping. If you don't catch a swarm, you won't be starting beekeeping until you do! I see swarm catching as a bonus way to get free bees!

Question 6: How much space do I need to raise bees?

Answer: People are raising bees everywhere! On farms, in backyards, and on balconies and rooftops! Most people have access to space for 1-2 hives where they live. If you don't you can probably find a local friend who would agree to having bees on their property in exchange for some free honey! Just make sure your bees are as close to you as possible, so you can take care of them properly! The more bees you want, the more space you need! Of course, keep in mind bees need access to water, nectar sources, and protection from wind. Some neighbors may not appreciate having bees near them, but this may be partially due to bee misconceptions.

Question 7: Can I make money as a beekeeper?

Answer: Yes! I made enough money my first year where I broke even on my investment, plus a little extra! Just remember, most businesses do not make profits for several years. Like any business, you need to invest to make a profit. You also need to grow sustainably, starting with one hive your first year and then suddenly jumping to 20 hives is probably not the best idea. There is a definite learning curve in beekeeping. Would you rather make a major mistake with one or two hives, or with twenty?

Question 8: Should I start with only 1 hive my first year?

Answer: If that's what your budget allows, then yes! If you can afford it, starting with two hives is usually preferred. As a first year beekeeper it is sometimes difficult to know what a "normal" hive looks like. By having two hives, you can compare your hives and you are better able to understand and see any potential problems. If you have a failing hive, you can sometimes combine that hive with your more productive hive.

Question 9: Is it a good idea to have a bee mentor?

Answer: Yes! A local bee mentor is best! Everyone does beekeeping a little differently, what works in one area may not work in your location. Working with a local mentor is a great way to learn if beekeeping is for you. Its nice to have a local friend to ask questions and to help you do some of your first inspections.

Question 10: What is killing the bees?

Answer: A combination of many factors. Disease, varroa mites, Colony Collapse Disorder, pesticides, climate change, fungicides, Nosema, the list goes on and on! The best strategy is better beekeeping and more beekeepers that are ready to try new strategies to keep the bees alive! This is my hope in designing this logbook, that we as a beekeeping community can keep better records and find ways to help save the bees!

THE LANGUAGE OF BEEKEEPING

Beekeeping has a language all its own! Start learning these basic beekeeping terms to start to understand the language of beekeeping! Remember, this is just to get started, there's so many more terms to learn!

Absconding – when an entire colony leaves its hive for various reasons, but basically the colony does not like the conditions of their home. One of the worst words in bee language. This term is different than swarming.

Apiary – the fancy term for a bee yard, derived From the scientific genus name of bees, Apis.

Apiarist- the fancy term for a beekeeper.

Bearding- when bees group together on the outside of the hive. Occurs on hot days to help keep the interior of the hive cool.

Bee Dance – a series of movements bees perform to communicate with other bees about locations of new food and potential new homes.

Bee Space – a space defined as between 6 mm and 8 mm, the space bees feel comfortable traveling through. Less than 6 mm and bees will close space with propolis. Larger than 8 mm and bees will build comb.

Brood – the egg, larva, and pupae of a bee colony. The bee babies!

Burr Comb – when bees build comb where the beekeeper doesn't want it. Usually a bee space issue.

Capped Honey – when bees have covered their honey storage for later use, important signal for harvest time!

Colony – all the members of your beehive, including brood, workers, drones, and the queen.

Cluster – when winter bees form a large ball of bees around the queen, usually in the winter time to stay warm.

Dearth – when there is no nectar available for bees to feed on. Beekeepers need to be aware of dearth to feed bees if needed.

Drawn comb – when worker bees have extended wax foundation into hexagonal shapes on the frame.

Drone – the male bees, main purpose is only to mate. Most are kicked out of hive in the late winter.

Flight path – the path bees take when leaving and entering the hive entrance.

Foundation – premade wax sheets that are placed inside frames to help get your bees started drawing comb.

Honey Bound – when there is limited space for brood to be laid and reared due to honey overtaking the laying area. Not good!

Royal Jelly – a substance made by worker bees and fed to larva for a few days and for much longer to larva that become a queen bee.

Stores – how much honey your bees have stored away for the winter time. Don't take too much!

Super – any hive box placed above the main hive body box.

Honey Flow – This is when the nectar flow is going strong and bees are making a lot of honey! Important to know, so you can help bees make more honey by adding more supers. Honey flows vary by region, according to plant blooms, but most areas have 1-3 honey flows per year.

Mating Flight – when a virgin queen must leave the hive to mate with several drones in the air.

Nectar- gathered from flowers, this is what bees turn into honey.

Nuc – a starter hive that contains a queen, all stages of brood, and worker bees. Usually 4-5 frames.

Orientation flights – flights young adult bees make before they become forager bees. Bees are learning landmarks to be able to find their hive and not get lost. These flights also occur when a hive is moved to a new location.

Propolis- a substance harvested from the resin of trees by bees and used to seal hive bodies and gaps.

Queen cell – a special cell drawn by worker bees which will house a worker bee larva that will be made into a future queen. When drawn out and finished, resembles a peanut.

Queenright – when a colony has a functioning, mated queen that is laying eggs.

Re-queen – when the beekeeper replaces a queen with a new, usually younger queen.

Robbing – when a healthy hive, or sometimes other insects, steal honey from a weak hive. Usually a sign of a larger problem or dearth.

Supersede – when worker bees make a queen cell, usually in the middle of a frame. Queen is either missing, old, or failing to produce enough eggs.

Swarming – occurs when a usually older queen leaves the hive with around half of the workers to start a new colony. Not a good day when your colony swarms.

BEEKEEPING SUPPLIES:

Keep track of all your beekeeping purchases, including bees, tools, equipment, wooden ware, and honey bottling supplies. Keep track of everything to compare your income to your expenses to see if your growing hobby has turned into a small business. Or, give all your honey away to friends!

Item Purchased	Price	Purchased From	Date Purchased

Item Purchased	Price	Purchased From	Date Purchased

Item Purchased	Price	Purchased From	Date Purchased

A Year of Blooms

Flowers! You probably didn't know this, but every beekeeper eventually becomes an amateur botanist as well. As a beekeeper, you will come to appreciate the flowers in your area and start noticing what flowers bloom, and when they bloom! For example, in the Blue Ridge Mountains of western North Carolina where I live, the first major bloom is tulip poplar, followed by many wildflowers and sourwood. In the fall I am lucky enough to have an amazing goldenrod bloom! Remember that flowers include flowering trees, and many plants that most people consider roadside weeds. We aren't just focused on ornamentals!

Why should you care about what flowers are blooming? First, that means nectar for your bees, which means you probably don't need to feed your bees when there is a nectar flow. A major nectar flow will signal to you that it is time to keep adding more supers, so the bees can harvest nectar and start making honey! If you have entered a nectar dearth, then you may want to give your bees supplemental food.

By keeping track of the major flower blooms, you can tell your customers what type of honey they are buying. You can also start to figure out what plants to plant on your own property. Use this chart below to keep track of the major flower blooms in your area. Most of us aren't going to be using the winter months, but if you are a Florida, deep Texas, or southern California beekeeper, you just might! Under each month I would not only list what major flowers are in bloom, but add terms for bloom strength, such as fair, good, and excellent to help you better describe the amount of blooms in your area. For example, I only have a few sourwoods near my bees, so I would only say I have a fair sourwood bloom. In the fall I have amazing goldenrod everywhere, so I would rank this as an excellent bloom. I would not usually sell my honey as sourwood, but in the fall, I confidently tell customers they are getting goldenrod honey, since no other flowers are blooming in early October! Here is a great website resource by NASA, yes NASA, to help you learn more about what typically blooms in your area: https://honeybeenet.gsfc.nasa.gov/Honeybees/Forage.htm

January Flowers in Bloom	February Flowers in Bloom
March Flowers in Bloom	**April Flowers in Bloom**
May Flowers in Bloom	**June Flowers in Bloom**

July Flowers in Bloom	August Flowers in Bloom
September Flowers in Bloom	October Flowers in Bloom
November Flowers in Bloom	December Flowers in Bloom

APIARY SETUP:

Where should you put your beehives and apiary? There are 3 factors you need to consider, the bees, the beekeeper, and everyone else that may be near the bee yard! Most beekeepers say placing your hive entrances east or south east encourages bees to start working earlier in the morning. Having light shade in the afternoon means your bees have less fanning to cool their hive. Having level dry ground that is semi protected from wind and has a water source nearby is helpful. Bees however are more flexible than you think, and many hives have thrived regardless of direction of the entrance or amount of shade. You need to start with the parameters of your property and find the best apiary placement that works!

Often when setting up your bee yard, you may forget about what you the beekeeper needs to do to help your bees flourish. Placing a hive against a fence or house may make sense to save space or block the wind, but when it comes time to do a bee inspection, having access to the back of the hive is crucial! You don't want to inspect your bees from the front of the hive, this is the worst place to be! If the back of your hive is blocked, you can't slide mite boards and pollen boards in and out from the back of the bottom board. Having your hive as close as possible to vehicle access is also a major bonus. When it comes time to harvest your honey, those supers will be heavy! Being able to place them in the back of a vehicle will make the workload of carrying the honey loaded supers easier and lower the aggressiveness of the bees trying to follow you and their honey away from the bee yard. If you place your hives to close to the entrance to your home, you may be dealing with bees gaining access to the indoors, or even stinging unsuspecting visitors.

Thinking about the rest of your family, neighbors, and folks that might be near your hive is another factor to consider. If you place your hives near a major traffic area of your property, or near a place people gather, then more bee stings will most likely occur. If you live in an urban area, think about nearby sidewalks and neighbors. It may even be illegal to keep bees in some city limits (if so, maybe you can help change that!). Some neighbors may love having bees nearby, others may strongly oppose your hobby! That doesn't mean your neighbor gets to decide if you keep bees, but it may be worth finding a solution that works for everyone.

DESIGN YOUR APIARY:

APIARY REDESIGN:

As your apiary grows, you may eventually need to redesign your apiary to fit more hives!

THE BEE LIFE CYCLE

(and how understanding it will help you with your hive inspection)

Biology time! Another one of the many aspects of beekeeping, understanding the life cycle of the various types of bees! By learning how your bees grow, you will have more knowledge that will aid you during your hive inspections.

There are three types of bees in your hive, the egg producing queen, the generally non-egg producing female worker bees, and the male drones, whose main purpose is to find a queen and mate. All bees start as eggs, hatch into larva, undergo metamorphosis as pupa, and then emerge from cells as adult bees. However, each bee type's journey from egg to adult is not the same! Below is a general overview of the time each bee type spends in each life stage:

Bee Type	Egg	Larva	Cell Capped	Emerges from Pupa	Fertile After
Queen	Up to 3 days	Up to 8.5 days	Day 7.5	16 days	Day 23
Worker	Up to 3 days	Up to 9 days	Day 9	21 days	
Drone	Up to 3 days	Up to 9.5 days	Day 10	24 days	Day 38

The Mating of the Queen

A virgin queen will take several nuptial flights and mate with drones from other colonies. Drones usually die after mating, having been successful in passing their genetic makeup to future generations. The queen then returns to her hive and begins laying. She will lay only a single egg in each cell, which looks like a small grain of rice. If she lays a fertilized egg, this egg will become a female worker bee, or perhaps a new future queen bee. If she lays an unfertilized egg, the egg is destined to become a male drone. A queen will typically live 2-4 years, but will become less productive overtime, which may signal to the worker bees to make a new queen!

Female Worker Bees:

Worker bees arise from fertilized eggs laid by the queen. Worker bees take 21 days from the time an egg is laid to the time they emerge from their capped cells into adult bees, where they will take on a variety of jobs and roles. An average worker bee lives about 6 weeks during the summer, but in early fall, "winter bees" are produced by the queen that can live for up to 5 months. These bees are not foraging much and are expending less energy overall, so they live longer.

Drones:

When the queen bee lays an unfertilized egg, a drone is born! Drones make up around 5 percent of the colony, but they are only born to mate. They do not forage or protect the hive, they don't even have a stinger. Drones live about 8 weeks, but in the fall, most are literally kicked out of the hive by the worker bees. In the winter, resources are scarce, the worker bees don't want another mouth to feed that isn't contributing to the hive! Whereas worker bees take 21 days to emerge from a capped cell, drones are significantly larger and take 24 days to grow from an egg into an adult male bee. Their capped cells are noticeably larger and usually laid towards the bottom of drawn comb.

Drone Cells: Notice how drone cells look a lot like corn pops the cereal. Drones are larger, so they need more room and time to grow. The worker bees make the initial cells wider in circumference and the convex cap gives the drone larva even more space to grow into adult male bees.

Worker Bees: Enjoying the fruits of their labor!

The Many Jobs of a Worker Bee:

The name worker bee really doesn't even begin to do these bees justice! What do worker bees do? The real question should be, what do worker bees not do? Worker bees keep the hive clean, remove the hive of dead bees, make special food to feed larva, young drones, and the queen, process nectar into honey, protect the hive entrance from invaders, regulate the hive temperature, and go out into the world to find nectar, pollen, propolis, and water. Below is an overview that shows what a typical worker bee will do during their lifetime. Note that some of these duties will overlap.

Day 1-3

These young bees will be responsible for keeping the hive clean, usually starting with cleaning their recently departed cell so that a new egg can be laid by the queen in the now empty real estate.

Days 3-16

These bees are the undertakers of the colony. Remember that in a large healthy hive, anywhere from 800 to 1200 bees a day are dying of natural causes in the spring, summer, and fall. If your queen is producing 1500 to 2000 eggs a day, your hive should be growing! Most of the dead bees die while out foraging, but some die inside the hive. The undertaker bees oversee taking these deceased bees out of the hive, sometimes a fair distance away, or sometimes right past the hive entrance, especially if there are not enough worker bees that are able to help with this task.

Days 4-12

The bees are now nurse bees, and in charge of feeding the growing larva, young drones, and queen larva. By the end of the first week, worker bees have developed glands on their heads and mouths that produce and secrete special brood food, including royal jelly.

Days 7-12

Some bees at this age will become what is known as the queen's retinue, a group of special attendant bees that attend to the queen's needs. They loyally follow the queen and clean, groom, and feed her when needed. When the queen defecates, they clean it up! By having attendants, the queen can focus on her main task of laying those 1500 to 2000 eggs a day in the spring and summertime!

Days 12- 18

Meet the honey making bees! These bees collect the nectar from the foragers and add special enzymes to begin to lower the water content of the nectar and to process it into honey. These bees are also the wax makers and are drawing comb, and capping honey. They also help regulate the hive temperature through the fanning of their wings. This requires a lot of energy, so even though they are making honey, they are consuming quite a bit as well.

Days 17-22

The guard bees! They are protecting the hive from predators, but also other bees that are attempting to rob the hive of honey stores.

Day 22- Death

At this point, bees have now become forager bees and its their job to find nectar. Most bees in the spring and summer only live about 6 weeks, in the winter, they can live about 4-5 months.

The Making of a Queen Bee:

Yes, that's right queen bees are made, not born as a queen! It's the worker bees that choose which female bee will become a queen. Queen bees start out exactly like every other fertilized egg that eventually becomes a worker bee. So, what happens during development that causes a worker bee to become a queen bee? It's all about the food! Both future queen bee larva and worker bee larva are fed royal jelly, a substance secreted from glands in the heads of adult worker bees. After the third day, only future queens are fed royal jelly exclusively, whereas the worker bee larva will be fed honey and pollen. Scientists are finding that the feeding of proteins found in royal jelly activates different genes in future queen bees than in worker bees, so it is just as important what developing queens aren't fed (honey and pollen), as what they are fed, royal jelly. After 16 days, the new queen emerges from her pupa.

Why do Worker Bees Make New Queens?

Worker bees will make a new queen for several reasons. First, and perhaps most obvious is the original queen has either died, or gone missing, but for whatever reason, the hive is now queenless. Second, the current queen may be getting less productive and older, so the worker bees start making queen cells to grow a new queen to supersede the existing queen. The last reason is that the hive is getting ready to swarm, or to split the hive in two (more on this later).

3 types of Queen Cells:

Queen cells are specially made wax cells to potentially house the development of a future queen bee. By paying attention to what type of cells are built, a beekeeper can better understand the state of their hive. The first type of queen cell is a swarm cell, which looks like an upside-down teacup until an egg is laid by the queen inside of it. After this occurs, the worker bees develop the cell so that it looks more like a small peanut. The location of the cell is almost always built on the bottom of frames, which is very important to keep in mind for your hive inspections. These cells can indicate that your hive is getting ready to swarm, or just that it is healthy and has plenty of stores and is building swarm cells in the event that the colony gets to large. Remember, swarming, although not what a beekeeper wants, is a natural process for honey bees, and is how bees increase their population and territory. The queen willfully lays an egg in a swarm cell, knowing that a new queen will me made by the worker bees. When a swarm occurs, the old queen usually leaves the hive with part of the colony, so she isn't laying her replacement that is going to battle her to the death. Instead laying an egg in a swarm cell is the queen's choice.

The second type of queen cell, a supersedure cell is not the queen's choice! A supersedure queen cell is made by the worker bees when they feel it is time to replace the existing queen. Since this is a worker bee decision, and not a queen bee decision, the worker bees must take an existing worker bee cell that already contains an egg or developing larva and build the queen cell around it. If you were a queen bee would you willingly lay an egg knowing it meant your impending death? The supersedure cell is usually located in the middle of a frame and lets you know your queen may be replaced. The last cell type is an emergency cell, this is also built on the side of a frame and looks much like a supersedure cell. The difference here is that a supersedure cell was planned out by the hive, whereas an emergency cell was not! Something happened to the queen, she died, never came back to the hive, or maybe you accidentally squashed her when doing an inspection (it happens).

Queen Cells: Shaped like a peanut and on the bottom of the frame! Sign that hive may be getting ready to swarm or needs to be split.

The Queen's Retinue: Find the queen! Notice her elongated abdomen and how the worker bees have formed a circle around her. The retinue will protect, groom, and feed the queen. Looking for the retinue can help you find the queen.

HOW TO FIND THE QUEEN

Finding the queen in your colony will probably be one of your biggest challenges as a new beekeeper, I know it was one of mine! A lot of new, and old beekeepers even, will choose to get a marked queen. This is a great way to easily identify the queen, just look for the bee with the painted thorax, usually with brightly colored paint. Paint colors are becoming standardized according to the year your queen was raised, so the paint not only gives you an unmistakable identification, but helps you know the age of your queen and if she is getting old and needs to be replaced. Not finding her also lets you know she has either perished or been replaced by a new queen.

A lot of us however, will not have a marked queen either because you don't know it is an option, marked queens are not available in your area, or you don't want to pay the higher price. So, to find the queen, we must rely on the physical differences of the queen verse worker and drone bees, the queen's behavior, and the behavior of the bees around the queen.

Physical Appearance of the Queen:

The queen is the longest bee in your hive, with a significantly longer abdomen than a worker bee. Her abdomen is tapered and more cylindrical than a worker bee. The end of a worker bee's abdomen is shaped more in a U pattern, whereas a queen bee's abdomen is more in a V shape. Her wings also appear shorter, since the abdomen is elongated. The legs of a queen bee are splayed outward and are much more visible than the legs of a worker or a drone. Although more difficult to see during an inspection, drones have no stinger or pollen baskets, workers have barbed stingers and pollen baskets, and a queen is lacking pollen baskets and has a smooth stinger that can be used to sting multiple times. Don't worry, a queen will rarely sting you, she is saving her stinger to battle other queens! A new beekeeper may mistake a drone for the queen, but drones have more of a barrel shape, a shorter abdomen, and their eyes are extremely large. A good way to make sure you are looking at a drone is to look around and find another drone! A well-established hive will typically have several hundred drones, but only one queen!

Queen Bee Behavior:

The queen has one main job, and that is to lay eggs. So, the best place to start your search for the queen is on a brood frame. Although possible, a queen will rarely be on a frame that is just honey. Look for uncapped brood with recently laid eggs. She will most likely be on an interior bottom frame of a vertical hive, or in the center of a horizontal hive. Initially look in the center of a frame, but the longer you inspect the frame, the more likely the queen will feel exposed and may retreat to a corner or the bottom of the frame, in an attempt to hide. Be careful that you keep your brood frames directly above your hive. If your queen falls from a frame, hopefully she will fall back into the hive, and not onto the ground where she could be stepped on. Another trick is to look for a bee that might appear to not be doing much, this could be your queen resting.

How the other Bees React to the Queen:

The other bees will tend to part for the queen and allow her to pass, and then cluster behind the space she was occupying. The queen is also escorted by her retinue, a group of bees that will be attending to the queen, and most likely feeding her. Check out the picture on the previous page to see a retinue. There is always the chance they are feeding a new virgin queen, or a young bee, but most likely you have found the queen! If a hive is missing a queen, or you remove a queen from a hive, the worker bees will in unison make a distinctive hive buzz. It really is incredible to hear but takes practice to recognize!

Queen spotting: Can you find the queen? One of the struggles for new beekeepers!

Marked Queen: Even though this is a black and white photo, you can still easily pick out the marked queen, which has a green dot on her thorax. Marked queens help new beekeepers find the queen easily and learn how the other bees react to the queen. Marked queens are also great for keepers with a lot of bees. If you can't find the marked queen, she is either hiding on a frame, or the hive is queen less!

WHEN YOU CAN'T FIND THE QUEEN

Do you need to always find your queen in the colony? No, not always. As a new beekeeper, finding the queen is a struggle. You are getting used to and comfortable with your bees, and at first it can be intimidating. You also must find one individual out of tens of thousands of bees, not an easy task. Having a marked queen can be extremely helpful in recognizing your queen and allow you to better observe her behavior and the behavior of the bees around her, however the paint on a marked queen can wear off over time, and there's always the chance that your worker bees will decide to make a new queen.

There will be times when you absolutely need to find your queen, such as when splitting your hive, your last few fall inspections, and your first few spring inspections. It is still very valuable to find your queen during most full inspections, but if you can't, here are some other signs to look for that will point to recent queen activity.

Basically, you need to look for different stages of brood, which are signs your queen has been recently active. Finding capped brood, or even young bees emerging, tells you your queen has been active at least 20-25 days ago, not a great measure, right? Observing larva tells you your queen was most likely active within the last week, since most larva is capped around days 8 and 9. Finding eggs is a sure sign that your queen has been active within 3 days, since eggs hatch into larva around this time. Most inspections, except in winter for most of us, you should be finding eggs. Of course, you need to be aware that a worker bee can lay eggs, usually due to the hive being queenless. Worker bees can only lay drone eggs, so if you suddenly have tons of drones, your hive may be queenless. With proper inspections, you should be able to notice this happening. Signs include multiple eggs laid in one cell, and eggs laid on the side of a cell, since worker bees' abdomens are not long enough to reach the bottom of a cell.

COMMON PESTS AND PARASITES OF HONEYBEES

Note: There are a lot of ways, techniques, and theories on how to treat various pests and parasites. Some have been researched and proven to work, some may just be old beekeeper lore passed down through generations of beekeepers. Some folks choose to do nothing, or very little when they see pests, instead they think better bee breeding is the answer. Others will try every treatment possible to keep their bees alive and hopefully thriving. This section is not going to tell you how to treat your bees, just how to recognize some common pests and parasites.

American Foulbrood:

A bacterial disease that affects bee larva and is usually fatal for the bee colony. Bacteria will contaminate food sources in the hive and be fed to larva, where they live and feed inside the gut, eventually leading to death. Signs include a spotty brood pattern, moisture on sealed brood, and oozing coming from brood cells. Sunken sealed cells are a sign that larva is decomposing inside the cell. Many states have a burn only policy of infected hives, very difficult to treat.

Bears:

Most of us are dealing with black bears, found in most states. Surprisingly I know more urban beekeepers than rural beekeepers who have had problems with bears. Bears near urban areas are already habituated to people and will very nonchalantly cruise right into your backyard. Rural people tend to be hunters, and this keeps the bear population lower and usually makes more wary bears. The best solution is an electric bear fence. Once bears find your hive, they will return!

Bear Attack: My friend's hive torn apart in the winter by a city bear. Notice the bees still in cluster. This hive survived!

Mice:

That's right mice! This is most often a problem during the winter time. Mice move inside your hive in the winter to stay sheltered, warm, and to have a steady supply of winter food, honey, bees, and larva! In the warmer times of the year, bees can keep mice away, but during the winter, bees are just trying to survive in their cluster and stay warm and often are defenseless to the mice invaders, which may only be a frame away munching on honey and larva. Signs include chewed on frames and mouse droppings.

Nosema:

A fungal infection that causes young bees to not digest food efficiently and to not produce food for young brood. Young infected bees tend to skip the nursing stage and become forager bees at a very early stage. Reduces the typical lifespan of a bee.

Small Hive Beetles:

These small beetles you can easily see with the naked eye and are about the size of a drone's eye. They can cause damage to comb, honey and pollen stores. If the population is large, can cause a hive to abscond.

Tracheal Mites:

You can't see these little parasites, they are microscopic! They infest the trachea of bees, eventually resulting in death. Signs include decreased flight efficiency, and a larger than average number of crawling bees that aren't flying.

Varroa Mites:

I hate these little buggers! They are visible with the naked eye and look like tiny little ticks. They are parasitic and use bees as their host. The mites attach to larva and suck blood and other nutrients from the bees. Can eventually over run a colony and cause the colony to collapse.

Wax Moths

Not a true attacker of bees, but instead they eat bee honeycomb and wax. A large whitish larva that grows to many times the size of a typical bee. Adults are gray white moths. Usually wax moths can be handled by a healthy bee colony but can invade improperly stored wax frames in the winter and utterly destroy the comb that the bees have worked so hard to build.

Yellow Jackets:

These predators usually appear in the late summer, when their typical food sources are depleted and harder to find. They not only are trying to get a taste of honey but will also attack and consume your bees. Can stress a hive enough to cause them to abscond.

A Year of Beekeeping

Here is an overview of a year of beekeeping, a guide to help you get started and to figure out what the heck you should be doing! Please note, this is just a guide! This schedule works for most of America, for those who have a typical four seasons, including a winter! If you live in higher latitudes, say most of New England or in states that border Canada and have longer winters, you may need to push this schedule back a month. Folks who live in more temperate places, like Florida, Texas, or southern California may need to fast forward the schedule a month. Unexpected weather conditions, such as drought or extreme rain will change your bees schedule and survival. Elevation can also play a role in your beekeeping schedule. I keep bees in the mountains of western North Carolina, and folks that live near the coast have the start of spring a few weeks earlier! Adjust this schedule as needed, talk with other local beekeepers, and first and foremost, spend enough time with your bees to know what they NEED!

January

1st year beekeepers:

If you want to start beekeeping, this is the time to order your bees! There are a lot of options out there, including online companies and local bee farms. Make sure you know what stock or subspecies of bees you are getting, how many pounds of bees, and if you are getting a marked queen. You can also decide if you want a bee package or a nuc!

Seasoned Beekeepers:

Only check on your bees on a warm day! 50 F and above is when I will open a hive (unless I think there is an emergency, like starvation). If your bees are clustered, you don't need to open your hives any further, keep those bees in cluster! If needed, add pollen supplements, bee candy, fondant. There aren't any drones in the hive, but if your lucky, the queen may have started laying worker brood. You may need to brush dead bees away from the entrance to allow access.

Combine weak hives and start to check your equipment. Repair what is needed and decide how many hives you want this year. Order bees, or plan on making splits or catching some swarms! If you have a big snow, make sure your bees have access to the outside and proper ventilation.

February

1st year beekeepers:

Order your hives and safety equipment. For hives you need to decide on 8 or 10 frame, deeps, mediums, or shallows. There's also the horizontal hive option, which is growing in popularity. For safety equipment, you need a bee veil, smoker, and hive tool at the bare minimum. I also have a bee jacket, frame gripper, and a frame holder that I normally use for most inspections.

Seasoned Beekeepers:

For most of us, we are still in winter! Bees are still in cluster, no major nectar flow, so hives may need additional feeding. 1-1 bee syrup can be used or bee candy. 1st major pollen flow of red maple is about to start; you may see your bees bringing in pollen. Again, combine hives if needed, only open hives on a warm day. Brood buildup should be much more apparent than January. Towards the end of February, you may want to add a super of wax foundation for the bees to begin drawing comb.

March

1st year beekeepers:

Hopefully you have all your tools and equipment! Go ahead and paint your hives if needed and if you know any local beekeepers, this is a great time of year to see if you can help with some inspections. Get your feet wet before your bees arrive. This time of year, hives are still building in population, so working with less bees is easier for new beekeepers. Usually the end of March is the earliest that bought bees are available and delivered, especially in the south.

Seasoned beekeepers:

Early season swarms are possible! Implement strategies to reduce swarming, including making splits, checkerboarding, adding more supers, or temporarily removing the queen. Set swarm traps to catch bees.

You may still be feeding your bees and you can assess if you need to treat your bees for pests. Most treatments need to be done before you add any supers you will harvest honey from. A lot of hives will have moved up to the 2nd super after consuming most of the honey stores below. This is a great time to just reverse the boxes so the bees are back on the bottom and will build upwards again.

Make sure your hive is queen right and she is laying! This time of year, the brood is starting to build up at a rapid pace. There is still not a lot of nectar in most areas, so be prepared to feed still. March is when a lot of hives starve and use up their remaining winter stores. Maple, Willow, Alder, and dandelions will be in bloom in most areas towards end of the month!

April

1st year beekeepers:

This is the month a lot of folks get their 1st bees! If you can, have a seasoned beekeeper help you with your first installation of your bees. If you ordered a package, the bees will need to eat through the candy tunnel that releases the queen from her queen cage. This usually takes about 3 days, after 3 days, release the queen if needed. Just make sure she is released into the hive! Add your entrance reducer, this gives your small hive less space to guard and helps protect the food stores from other hives and robbers. Begin feeding 1-1 bee syrup, which is essential to help your bees build up valuable drawn comb. The queen can't lay if the worker bees haven't made enough cells. Be careful about over feeding, which can lead to your hive becoming "honey bound", or in this case, sugar water bound. Only put in enough bee syrup for a few days, and then wait a day or so to feed more. After your queen mates, you should start seeing eggs!

Seasoned Beekeepers:

Swarming is still possible, be prepared! Add more supers if needed for hopefully a good nectar flow in May. Give the bees time to draw out comb if needed. Your hive should still be expanding, queen should be laying, adult drones should be easy to find, and worker bees should be bringing in nectar and pollen. If your main goal is to make bees and not honey, this is a great time to make splits. The end of feeding your bees for most keepers. Some keepers will treat for mites and other pests now.

May

1st year beekeepers:

For most areas, probably the latest time of year to start a new colony or nuc. Hives started earlier in the year should have at least the hive body frames drawn out. If you haven't added a super, this is the time! In general, is your hive body or super has 7-8 frames drawn, you should add the next super. You may still be feeding your bees 1-1 syrup. Remember, you aren't harvesting honey anytime soon, with only the possibility of an early fall harvest (but don't count on it). Your first year is all about letting the bees draw comb for this year and then you can reuse the comb for several years. Great time to get comfortable looking for your queen and identifying eggs, larva, brood, and drones.

Seasoned beekeepers:

Most areas will have their first big nectar flow. Where I live, I'm hoping for black locust, tulip poplar, cherry, and blackberry to start blooming. Pay attention that you give your bees supers when needed and use queen excluders if desired. Limit inspections, since over inspecting can decrease honey harvest. Still a good time to make splits and find swarms. Southern latitude keepers may even have their first harvest of honey.

June

1st year beekeepers:

You should have a hive body drawn out and a 1st super that is drawn. This is honey for your bees. This is what your bees need to get through the winter. If you add another super, bees may draw the comb for next year. Use your queen excluder to keep the queen from laying in the extra honey super. If your hive is booming, can probably remove the entrance reducer.

Seasoned beekeepers:

For most beekeepers, this is harvest time and then the start of a dearth. Northern beekeepers, this is probably when your area is starting to see its first major flow. After harvesting, replace wet supers back on hives. You may want to put entrance reducers back on to reduce robbing. Check on queen health, she will still be laying, but may start to see a slight decline. Some folks will feed during this dearth to boost hives, but I prefer to wait.

July

1st year beekeepers:

Hive should be booming. Check on queen health, make sure you can find her. If you can't, call a friend to find her. No queen means no bees, unless your hives make a new queen!

Seasoned beekeepers:

Still a dearth for some, for others you may have the end of a flow happening. If your lucky, you may have species like basswood and sourwood that are a great end of spring/early summer flow. Pay attention to see if you need to add supers. Check for mites (this is good to do all year anyways), usually this time of year is when mites are at a maximum.

August

1st year beekeepers:

Hopefully you have an established hive body and super with honey. If you added a 2nd super, you may have extra drawn frames, even a little honey to harvest. Queen will start to decline her laying and you may see a decrease in population starting. Treat for mites! Add entrance reducers to minimize robbing, some keepers begin supplemental feeding, not recommended if you are trying to get a small fall harvest. Just remember the bees come first!

Seasoned beekeepers:

For some this may be the time of your first harvest (later in the month), or for a 2nd harvest (southern beekeepers). Dearth time! Most keepers treat for mites this month. Queen will start to decline in her laying. For most beekeepers, this is the latest you would make a split or a nuc colony. Some beekeepers will start feeding 1-1 syrup to boost colony size. Start thinking about combining weak hives if needed and adding entrance reducers to prevent robbing. Yellow jackets are out in full force during this dearth.

September

1st year beekeepers:

Getting ready for winter! You may notice the worker bees starting to get the drones from the hive. A fall bloom of asters, goldenrod and other flowers may allow you a small harvest. Remember this is not your goal, just if you are lucky. The biggest mistake new beekeepers make is getting greedy and taking honey from their bees! 50 % of the time I don't harvest from a 1st year hive. The goal your first year is making comb and having your bees overwinter successfully! Begin supplemental feeding if you haven't already.

Seasoned beekeepers:

Getting ready for winter! A fall bloom in some areas may even mean a small harvest. Just make sure your bees have 45-60 lbs. of honey for the winter. I have a great fall flow, so I wait until now to start supplemental feeding. 1-1 syrup encourages more brood rearing and 2-1 encourages more syrup storage. Don't forget pollen supplements.

October

All Beekeepers:

Are your bees alive? The next two months is when you see a lot of hives crash due to mites! Check on your queen, keep feeding. Remove extra supers and make sure there is honey for your winter bees. Hopefully your bees fill at least 5 frames going into winter (minimum). Switch to 2-1 syrup. If desired, add quilt boxes to keep moisture out.

November

All beekeepers:

Your bees are in cluster, don't open the hive unless you need to. Add bee candy or fondant to feed. Hopefully you left enough honey and your hive was queen right. If you have a queen problem, not much you can do about it now. Read this book by the woodstove.

December

All beekeepers:

December is a lot like November for beekeepers. Just colder and darker. Great time to plan and dream for the spring!

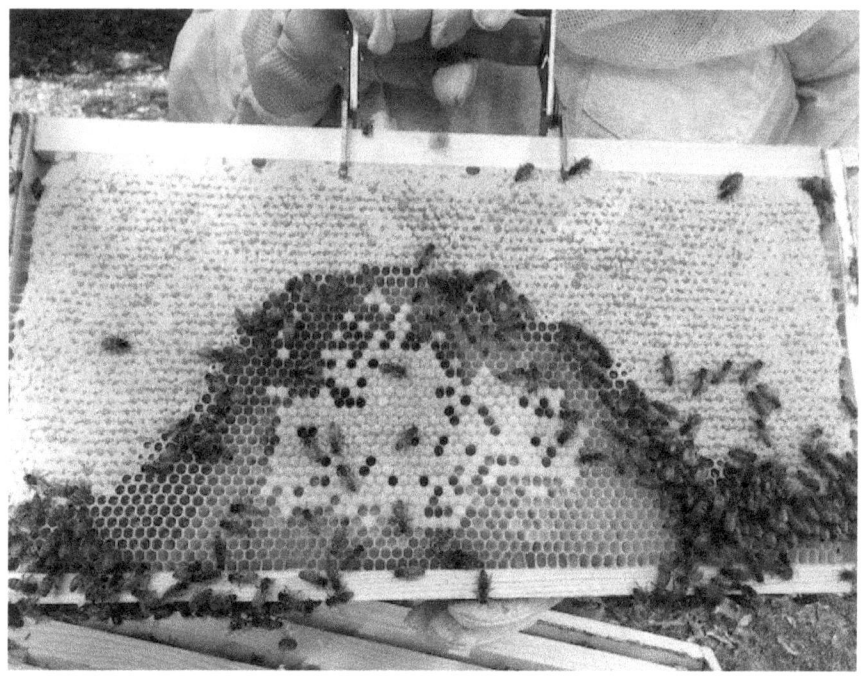

Hive Body Frame: Typical deep frame. Caped brood, open larva cells, and eggs are in the interior. Capped honey above and on the sides.

Bee Candy: Late winter bees enjoying sugar candy and a pollen patty.

HIVE INSPECTION LOG:

(What to look for during a hive inspection and how to use the inspection log)

The hive inspection! This is the whole reason I made this logbook, to become a more conscious and thorough beekeeper. At the end of my first year as a beekeeper, I lost both of my hives, one to mites, and the second most likely to a weak queen and moisture problems. At the end of my first year, I realized that to become a better beekeeper, I needed to learn and be aware of what problems and factors are affecting my bees. If you don't even know what the potential problems might be, how can you ever begin to help your bees thrive?

To that end, I dug in and researched and talked to other beekeepers. Although I have learned every beekeeper has their own way of working their bees, and a bee inspection will vary according to the time of year, I found that there are certain major checks you need to make during your hive inspection:

I have updated and improved my log, so that 1 page is used for only 1 hive. This way, you can see the data of an individual hive over time, without having to flip pages. What I would do is use both the front and back pages for 1 single hive. This gives you 12 inspections throughout the year. If you want, skip the next page, and leave it available for future inspections for that hive as well, giving you room for 24 inspections per hive.

Here's how to use the hive inspection log:

Hive Name/Number/Letter:

If you have more than a few hives, I suggest labeling your hives, or at least making sure they are labelled on your apiary map. This way you can keep track of each hive. Especially if you try different methods on each hive to see what works best for you.

Hive Origin/Ancestry:

Record the history of each of your hives. Where did the hive come from? Was it bought as a package or nuc from a large bee farm, or local beekeeper? Was it caught as a swarm, or maybe it was a split from an overwintered hive? If you add a new queen, remember to change your hive ancestry. Over time, you may start to see what bee lines are most successful in your apiary.

Date and Weather/Temperature (W/T):

Record these important basics. A lot of beekeeping management is based on the bee life cycle. Having accurate dates can be crucial to understand the dynamics of your hive overtime and to plan for what you should be seeing in future inspections.

Entrance Activity:

Take a moment to observe your hive entrances for each of your hives. This is the first thing I do! A quick check on each of my hives can tell me a lot. Are they bringing in pollen? Is there a hive with minimal activity when all other hives seem to be active? If so, you probably should open that hive first and try to figure out why!

Drawn Frames and Frames of Bees

If you are adding fresh undrawn wax foundation, it's always good to monitor how quickly your bees are drawing out the wax and building comb for the queen to lay and to store honey and pollen. Recording an estimate of how many frames of bees your hive has really helps see the population change of your hive over the season. It can help identify if it is time to split, or if your hive has any major problems.

Eggs/Open Brood/Capped Brood (E/O/C)

Every inspection in the spring, summer, and fall should include any observations of eggs, open brood, and capped brood. You may not always be able to find the queen, however finding queen evidence can help you know she is there, or how long ago she was last laying. Looking at the laying pattern can help you determine if it is time to requeen, add more space, or split your hive.

N/F/G/A

N = no brood at this stage was observed

F = few brood at this stage observed

G= a good amount of brood at this stage observed, usually several frames

A= amazing! Lots of solid patterns at this stage of brood. Most likely seen in capped brood since it is the longest stage and lots of days will overlap.

As an example, let us say I start with a package, my initial observation of brood may be recorded as:

G/F/N = good number of eggs, few open brood, no capped brood

After a few weeks I would hope to see:

G/G/A = decent number of eggs and open brood and lots of capped brood!

In late fall/early winter, I may see:

F/F/F = showing that my queen has slowed down brood production due to seasonal factors

Queen Found:

Remember, you don't always have to find the queen in every inspection. Sometimes I'm happy just finding eggs or open brood, but if I'm planning on doing a split or a nuc, or during certain times of year I do my best to find the queen. I always find the queen in early spring and early fall, to make sure my hive is queenright.

Queen Cells

Queen cells can help you know what your hive is planning to do next. If you find them, you can implement certain strategies to benefit your apiary. Finding queen cells always mean there is a new queen coming soon! Typically, queen cells located on the bottom of frames indicate the hive is getting ready to swarm. The old queen usually leaves with about half of the hive to find a new place to live. This gives the old queen a chance to extend her life a bit and increase the time that the hive needs to rely on her. Queen cells located in the middle of the frames usually indicates that the hive is trying to supersede the current queen, or the old queen has perished, and the hive has made emergency queen cells. Hives generally make practice queen cups year-round, but a queen cell looks more like a peanut. You can even note if there is queen larva inside the uncapped queen cells. You can also use extra queen cells to make hive splits.

Mites and Other Pests:

Always be vigilant for mites and other pests! Do you notice anything unusual, any signs and symptoms that may point to a major hive problem? Mites, hive beetles, wax moths, and other pests can damage your bees, foundation, and honey.

Supplemental Feed and Treatments:

Some beekeepers do not believe in supplemental feeding, but if you do, here is the spot to mark what type of supplemental feed you are giving your bees. Sugar water, bee candy, fondant, pollen patties, whatever you think your bees need, it's up to you to decide. Using the bloom chart in the front of the book will be helpful to help you be aware of seasonal dearths in your area. Treatments for pests can also be noted here.

Honey Frames u/c:

Certain times of year you may be overflowing in honey when a nectar flow is on, other times you may have so little that the bees may need some extra help. Keeping track of your uncapped and capped honey can help you know when it is time to harvest!

General Health and Actions Taken:

This section is to use however you want! Did you learn a bee lesson? Maybe you made a major mistake? If you decide to treat your bees for certain pests, you can record what you used and when the treatment will be done. You can also record if it was successful or not! Did you split a hive? Requeen? Do you see your bees bringing in pollen? Record anything you need to in this section!

How do your bees look in general? Are they out foraging, are the numbers high, do you have too many drones? Really look at your bees and make sure the colony seems healthy. Look for signs of distress that could be indicative of a dearth, pests, or weather-related problems. You can measure your colony using the frame method, where you mark how many frames of your hive are covered in bees. Overtime, you will be able to see if your hive is growing, or declining, which is natural towards the end of fall. Monitoring the colony size can help you know if your colony is going to swarm, or if it is time to split your hive.

General Overview of All Inspections and Hive Progression:

There are 6 inspections per page. After you have completed 6 inspections for the hive, there is a little space to write your general overall thoughts about the progression and status of hive. What's next for this hive? What worked for you? What will you change?

How Often Should You Do an Inspection?

This varies according to time of year, location, status of the hive, and your own preference. Establishing a new hive with no drawn comb, you may be inspecting weekly. In the summer, you might only inspect every 3 weeks, and in the winter, you may not open the hive completely for several months. Mite checks with pull out mite boards can be done weekly, without ever opening the hive. Its always best to be around your bees, even if you aren't doing a full inspection.

Understanding bee biology and the blooms in your area will help you significantly to determine when you do hive inspections.

| HIVE NAME/NUMBER/LETTER: |
| HIVE ORIGIN/ ANCESTRY: |

Date	W/T	Entrance Activity	Drawn Frames and Frames of Bees	E/O/C n/f/g/a	Queen Found (y/n)	Queen Cells (y/n) (where)	Mites & Pests found	Supp Feed & Treatments	Honey Frames u/c	General Health and Actions Taken

General Overview of All Inspections and Hive Progression:

HIVE NAME/NUMBER/LETTER:

HIVE ORIGIN/ ANCESTRY:

Date	W/T	Entrance Activity	Drawn Frames and Frames of Bees	E/O/C n/f/g/a	Queen Found (y/n)	Queen Cells (y/n) (where)	Mites & Pests found	Supp Feed & Treatments	Honey Frames u/c	General Health and Actions Taken

General Overview of All Inspections and Hive Progression:

HIVE NAME/NUMBER/LETTER:										
HIVE ORIGIN/ ANCESTRY:										
Date	W/T	Entrance Activity	Drawn Frames and Frames of Bees	E/O/C n/f/g/a	Queen Found (y/n)	Queen Cells (y/n) (where)	Mites & Pests found	Supp Feed & Treatments	Honey Frames u/c	General Health and Actions Taken

General Overview of All Inspections and Hive Progression:

HIVE NAME/NUMBER/LETTER:
HIVE ORIGIN/ ANCESTRY:

Date	W/T	Entrance Activity	Drawn Frames and Frames of Bees	E/O/C n/f/g/a	Queen Found (y/n)	Queen Cells (y/n) (where)	Mites & Pests found	Supp Feed & Treatments	Honey Frames u/c	General Health and Actions Taken

General Overview of All Inspections and Hive Progression:

HIVE NAME/NUMBER/LETTER:
HIVE ORIGIN/ ANCESTRY:

Date	W/T	Entrance Activity	Drawn Frames and Frames of Bees	E/O/C n/f/g/a	Queen Found (y/n)	Queen Cells (y/n) (where)	Mites & Pests found	Supp Feed & Treatments	Honey Frames u/c	General Health and Actions Taken

General Overview of All Inspections and Hive Progression:

HIVE NAME/NUMBER/LETTER:
HIVE ORIGIN/ ANCESTRY:

Date	W/T	Entrance Activity	Drawn Frames and Frames of Bees	E/O/C n/f/g/a	Queen Found (y/n)	Queen Cells (y/n) (where)	Mites & Pests found	Supp Feed & Treatments	Honey Frames u/c	General Health and Actions Taken

General Overview of All Inspections and Hive Progression:

HIVE NAME/NUMBER/LETTER:
HIVE ORIGIN/ ANCESTRY:

Date	W/T	Entrance Activity	Drawn Frames and Frames of Bees	E/O/C n/f/g/a	Queen Found (y/n)	Queen Cells (y/n) (where)	Mites & Pests found	Supp Feed & Treatments	Honey Frames u/c	General Health and Actions Taken

General Overview of All Inspections and Hive Progression:

HIVE NAME/NUMBER/LETTER:
HIVE ORIGIN/ ANCESTRY:

Date	W/T	Entrance Activity	Drawn Frames and Frames of Bees	E/O/C n/f/g/a	Queen Found (y/n)	Queen Cells (y/n) (where)	Mites & Pests found	Supp Feed & Treatments	Honey Frames u/c	General Health and Actions Taken

General Overview of All Inspections and Hive Progression:

HIVE NAME/NUMBER/LETTER:
HIVE ORIGIN/ ANCESTRY:

Date	W/T	Entrance Activity	Drawn Frames and Frames of Bees	E/O/C n/f/g/a	Queen Found (y/n)	Queen Cells (y/n) (where)	Mites & Pests found	Supp Feed & Treatments	Honey Frames u/c	General Health and Actions Taken

General Overview of All Inspections and Hive Progression:

HIVE NAME/NUMBER/LETTER:

HIVE ORIGIN/ ANCESTRY:

Date	W/T	Entrance Activity	Drawn Frames and Frames of Bees	E/O/C n/f/g/a	Queen Found (y/n)	Queen Cells (y/n) (where)	Mites & Pests found	Supp Feed & Treatments	Honey Frames u/c	General Health and Actions Taken

General Overview of All Inspections and Hive Progression:

HIVE NAME/NUMBER/LETTER:
HIVE ORIGIN/ ANCESTRY:

Date	W/T	Entrance Activity	Drawn Frames and Frames of Bees	E/O/C n/f/g/a	Queen Found (y/n)	Queen Cells (y/n) (where)	Mites & Pests found	Supp Feed & Treatments	Honey Frames u/c	General Health and Actions Taken

General Overview of All Inspections and Hive Progression:

HIVE NAME/NUMBER/LETTER:
HIVE ORIGIN/ ANCESTRY:

Date	W/T	Entrance Activity	Drawn Frames and Frames of Bees	E/O/C n/f/g/a	Queen Found (y/n)	Queen Cells (y/n) (where)	Mites & Pests found	Supp Feed & Treatments	Honey Frames u/c	General Health and Actions Taken

General Overview of All Inspections and Hive Progression:

HIVE NAME/NUMBER/LETTER:
HIVE ORIGIN/ ANCESTRY:

Date	W/T	Entrance Activity	Drawn Frames and Frames of Bees	E/O/C n/f/g/a	Queen Found (y/n)	Queen Cells (y/n) (where)	Mites & Pests found	Supp Feed & Treatments	Honey Frames u/c	General Health and Actions Taken

General Overview of All Inspections and Hive Progression:

HIVE NAME/NUMBER/LETTER:
HIVE ORIGIN/ ANCESTRY:

Date	W/T	Entrance Activity	Drawn Frames and Frames of Bees	E/O/C n/f/g/a	Queen Found (y/n)	Queen Cells (y/n) (where)	Mites & Pests found	Supp Feed & Treatments	Honey Frames u/c	General Health and Actions Taken

General Overview of All Inspections and Hive Progression:

HIVE NAME/NUMBER/LETTER:
HIVE ORIGIN/ ANCESTRY:

Date	W/T	Entrance Activity	Drawn Frames and Frames of Bees	E/O/C n/f/g/a	Queen Found (y/n)	Queen Cells (y/n) (where)	Mites & Pests found	Supp Feed & Treatments	Honey Frames u/c	General Health and Actions Taken

General Overview of All Inspections and Hive Progression:

HIVE NAME/NUMBER/LETTER:
HIVE ORIGIN/ ANCESTRY:

Date	W/T	Entrance Activity	Drawn Frames and Frames of Bees	E/O/C n/f/g/a	Queen Found (y/n)	Queen Cells (y/n) (where)	Mites & Pests found	Supp Feed & Treatments	Honey Frames u/c	General Health and Actions Taken

General Overview of All Inspections and Hive Progression:

HIVE NAME/NUMBER/LETTER:
HIVE ORIGIN/ ANCESTRY:

Date	W/T	Entrance Activity	Drawn Frames and Frames of Bees	E/O/C n/f/g/a	Queen Found (y/n)	Queen Cells (y/n) (where)	Mites & Pests found	Supp Feed & Treatments	Honey Frames u/c	General Health and Actions Taken

General Overview of All Inspections and Hive Progression:

HIVE NAME/NUMBER/LETTER:
HIVE ORIGIN/ ANCESTRY:

Date	W/T	Entrance Activity	Drawn Frames and Frames of Bees	E/O/C n/f/g/a	Queen Found (y/n)	Queen Cells (y/n) (where)	Mites & Pests found	Supp Feed & Treatments	Honey Frames u/c	General Health and Actions Taken

General Overview of All Inspections and Hive Progression:

| HIVE NAME/NUMBER/LETTER: |
| HIVE ORIGIN/ ANCESTRY: |

Date	W/T	Entrance Activity	Drawn Frames and Frames of Bees	E/O/C n/f/g/a	Queen Found (y/n)	Queen Cells (y/n) (where)	Mites & Pests found	Supp Feed & Treatments	Honey Frames u/c	General Health and Actions Taken

General Overview of All Inspections and Hive Progression:

HIVE NAME/NUMBER/LETTER:
HIVE ORIGIN/ ANCESTRY:

Date	W/T	Entrance Activity	Drawn Frames and Frames of Bees	E/O/C n/f/g/a	Queen Found (y/n)	Queen Cells (y/n) (where)	Mites & Pests found	Supp Feed & Treatments	Honey Frames u/c	General Health and Actions Taken

General Overview of All Inspections and Hive Progression:

HIVE NAME/NUMBER/LETTER:										
HIVE ORIGIN/ ANCESTRY:										
Date	W/T	Entrance Activity	Drawn Frames and Frames of Bees	E/O/C n/f/g/a	Queen Found (y/n)	Queen Cells (y/n) (where)	Mites & Pests found	Supp Feed & Treatments	Honey Frames u/c	General Health and Actions Taken

General Overview of All Inspections and Hive Progression:

HIVE NAME/NUMBER/LETTER:
HIVE ORIGIN/ ANCESTRY:

Date	W/T	Entrance Activity	Drawn Frames and Frames of Bees	E/O/C n/f/g/a	Queen Found (y/n)	Queen Cells (y/n) (where)	Mites & Pests found	Supp Feed & Treatments	Honey Frames u/c	General Health and Actions Taken

General Overview of All Inspections and Hive Progression:

| HIVE NAME/NUMBER/LETTER: |
| HIVE ORIGIN/ ANCESTRY: |

Date	W/T	Entrance Activity	Drawn Frames and Frames of Bees	E/O/C n/f/g/a	Queen Found (y/n)	Queen Cells (y/n) (where)	Mites & Pests found	Supp Feed & Treatments	Honey Frames u/c	General Health and Actions Taken

General Overview of All Inspections and Hive Progression:

HIVE NAME/NUMBER/LETTER:
HIVE ORIGIN/ ANCESTRY:

Date	W/T	Entrance Activity	Drawn Frames and Frames of Bees	E/O/C n/f/g/a	Queen Found (y/n)	Queen Cells (y/n) (where)	Mites & Pests found	Supp Feed & Treatments	Honey Frames u/c	General Health and Actions Taken

General Overview of All Inspections and Hive Progression:

HIVE NAME/NUMBER/LETTER:
HIVE ORIGIN/ ANCESTRY:

Date	W/T	Entrance Activity	Drawn Frames and Frames of Bees	E/O/C n/f/g/a	Queen Found (y/n)	Queen Cells (y/n) (where)	Mites & Pests found	Supp Feed & Treatments	Honey Frames u/c	General Health and Actions Taken

General Overview of All Inspections and Hive Progression:

HIVE NAME/NUMBER/LETTER:
HIVE ORIGIN/ ANCESTRY:

Date	W/T	Entrance Activity	Drawn Frames and Frames of Bees	E/O/C n/f/g/a	Queen Found (y/n)	Queen Cells (y/n) (where)	Mites & Pests found	Supp Feed & Treatments	Honey Frames u/c	General Health and Actions Taken

General Overview of All Inspections and Hive Progression:

HIVE NAME/NUMBER/LETTER:
HIVE ORIGIN/ ANCESTRY:

Date	W/T	Entrance Activity	Drawn Frames and Frames of Bees	E/O/C n/f/g/a	Queen Found (y/n)	Queen Cells (y/n) (where)	Mites & Pests found	Supp Feed & Treatments	Honey Frames u/c	General Health and Actions Taken

General Overview of All Inspections and Hive Progression:

HIVE NAME/NUMBER/LETTER:
HIVE ORIGIN/ ANCESTRY:

Date	W/T	Entrance Activity	Drawn Frames and Frames of Bees	E/O/C n/f/g/a	Queen Found (y/n)	Queen Cells (y/n) (where)	Mites & Pests found	Supp Feed & Treatments	Honey Frames u/c	General Health and Actions Taken

General Overview of All Inspections and Hive Progression:

HIVE NAME/NUMBER/LETTER:
HIVE ORIGIN/ ANCESTRY:

Date	W/T	Entrance Activity	Drawn Frames and Frames of Bees	E/O/C n/f/g/a	Queen Found (y/n)	Queen Cells (y/n) (where)	Mites & Pests found	Supp Feed & Treatments	Honey Frames u/c	General Health and Actions Taken

General Overview of All Inspections and Hive Progression:

HIVE NAME/NUMBER/LETTER:
HIVE ORIGIN/ ANCESTRY:

Date	W/T	Entrance Activity	Drawn Frames and Frames of Bees	E/O/C n/f/g/a	Queen Found (y/n)	Queen Cells (y/n) (where)	Mites & Pests found	Supp Feed & Treatments	Honey Frames u/c	General Health and Actions Taken

General Overview of All Inspections and Hive Progression:

HIVE NAME/NUMBER/LETTER:
HIVE ORIGIN/ ANCESTRY:

Date	W/T	Entrance Activity	Drawn Frames and Frames of Bees	E/O/C n/f/g/a	Queen Found (y/n)	Queen Cells (y/n) (where)	Mites & Pests found	Supp Feed & Treatments	Honey Frames u/c	General Health and Actions Taken

General Overview of All Inspections and Hive Progression:

HIVE NAME/NUMBER/LETTER:
HIVE ORIGIN/ ANCESTRY:

Date	W/T	Entrance Activity	Drawn Frames and Frames of Bees	E/O/C n/f/g/a			Queen Found (y/n)	Queen Cells (y/n) (where)	Mites & Pests found	Supp Feed & Treatments	Honey Frames u/c	General Health and Actions Taken

General Overview of All Inspections and Hive Progression:

HIVE NAME/NUMBER/LETTER:
HIVE ORIGIN/ ANCESTRY:

Date	W/T	Entrance Activity	Drawn Frames and Frames of Bees	E/O/C n/f/g/a	Queen Found (y/n)	Queen Cells (y/n) (where)	Mites & Pests found	Supp Feed & Treatments	Honey Frames u/c	General Health and Actions Taken

General Overview of All Inspections and Hive Progression:

HIVE NAME/NUMBER/LETTER:
HIVE ORIGIN/ ANCESTRY:

Date	W/T	Entrance Activity	Drawn Frames and Frames of Bees	E/O/C n/f/g/a	Queen Found (y/n)	Queen Cells (y/n) (where)	Mites & Pests found	Supp Feed & Treatments	Honey Frames u/c	General Health and Actions Taken

General Overview of All Inspections and Hive Progression:

HIVE NAME/NUMBER/LETTER:
HIVE ORIGIN/ ANCESTRY:

Date	W/T	Entrance Activity	Drawn Frames and Frames of Bees	E/O/C n/f/g/a	Queen Found (y/n)	Queen Cells (y/n) (where)	Mites & Pests found	Supp Feed & Treatments	Honey Frames u/c	General Health and Actions Taken

General Overview of All Inspections and Hive Progression:

HIVE NAME/NUMBER/LETTER:
HIVE ORIGIN/ ANCESTRY:

Date	W/T	Entrance Activity	Drawn Frames and Frames of Bees	E/O/C n/f/g/a	Queen Found (y/n)	Queen Cells (y/n) (where)	Mites & Pests found	Supp Feed & Treatments	Honey Frames u/c	General Health and Actions Taken

General Overview of All Inspections and Hive Progression:

HIVE NAME/NUMBER/LETTER:
HIVE ORIGIN/ ANCESTRY:

Date	W/T	Entrance Activity	Drawn Frames and Frames of Bees	E/O/C n/f/g/a	Queen Found (y/n)	Queen Cells (y/n) (where)	Mites & Pests found	Supp Feed & Treatments	Honey Frames u/c	General Health and Actions Taken

General Overview of All Inspections and Hive Progression:

HIVE NAME/NUMBER/LETTER:
HIVE ORIGIN/ ANCESTRY:

Date	W/T	Entrance Activity	Drawn Frames and Frames of Bees	E/O/C n/f/g/a	Queen Found (y/n)	Queen Cells (y/n) (where)	Mites & Pests found	Supp Feed & Treatments	Honey Frames u/c	General Health and Actions Taken

General Overview of All Inspections and Hive Progression:

HIVE NAME/NUMBER/LETTER:
HIVE ORIGIN/ ANCESTRY:

Date	W/T	Entrance Activity	Drawn Frames and Frames of Bees	E/O/C n/f/g/a	Queen Found (y/n)	Queen Cells (y/n) (where)	Mites & Pests found	Supp Feed & Treatments	Honey Frames u/c	General Health and Actions Taken

General Overview of All Inspections and Hive Progression:

HIVE NAME/NUMBER/LETTER:
HIVE ORIGIN/ ANCESTRY:

Date	W/T	Entrance Activity	Drawn Frames and Frames of Bees	E/O/C n/f/g/a	Queen Found (y/n)	Queen Cells (y/n) (where)	Mites & Pests found	Supp Feed & Treatments	Honey Frames u/c	General Health and Actions Taken

General Overview of All Inspections and Hive Progression:

HIVE NAME/NUMBER/LETTER:
HIVE ORIGIN/ ANCESTRY:

Date	W/T	Entrance Activity	Drawn Frames and Frames of Bees	E/O/C n/f/g/a	Queen Found (y/n)	Queen Cells (y/n) (where)	Mites & Pests found	Supp Feed & Treatments	Honey Frames u/c	General Health and Actions Taken

General Overview of All Inspections and Hive Progression:

HIVE NAME/NUMBER/LETTER:										
HIVE ORIGIN/ ANCESTRY:										
Date	W/T	Entrance Activity	Drawn Frames and Frames of Bees	E/O/C n/f/g/a	Queen Found (y/n)	Queen Cells (y/n) (where)	Mites & Pests found	Supp Feed & Treatments	Honey Frames u/c	General Health and Actions Taken

General Overview of All Inspections and Hive Progression:

HIVE NAME/NUMBER/LETTER:
HIVE ORIGIN/ ANCESTRY:

Date	W/T	Entrance Activity	Drawn Frames and Frames of Bees	E/O/C n/f/g/a	Queen Found (y/n)	Queen Cells (y/n) (where)	Mites & Pests found	Supp Feed & Treatments	Honey Frames u/c	General Health and Actions Taken

General Overview of All Inspections and Hive Progression:

HIVE NAME/NUMBER/LETTER:
HIVE ORIGIN/ ANCESTRY:

Date	W/T	Entrance Activity	Drawn Frames and Frames of Bees	E/O/C n/f/g/a	Queen Found (y/n)	Queen Cells (y/n) (where)	Mites & Pests found	Supp Feed & Treatments	Honey Frames u/c	General Health and Actions Taken

General Overview of All Inspections and Hive Progression:

HARVEST LOG

You did it! You were able to make your bees happy and the colony is hopefully thriving! Now, you can keep track of the bee produced goods your beehives provide! Of course, honey is number 1 to most folks, but once you develop the basics of beekeeping you can harvest a variety of goods, including beeswax, propolis, pollen, and honey. You may even start rearing queens or be able to sell a limited number of bees through splits! Whatever you do, have fun doing it!

Product Harvested: Date: Amount: From Hive(s):	Product Harvested: Date: Amount: From Hive(s):
Product Harvested: Date: Amount: From Hive(s):	Product Harvested: Date: Amount: From Hive(s):
Product Harvested: Date: Amount: From Hive(s):	Product Harvested: Date: Amount: From Hive(s):
Product Harvested: Date: Amount: From Hive(s):	Product Harvested: Date: Amount: From Hive(s):
Product Harvested: Date: Amount: From Hive(s):	Product Harvested: Date: Amount: From Hive(s):

Product Harvested: Date: Amount: From Hive(s):	Product Harvested: Date: Amount: From Hive(s):
Product Harvested: Date: Amount: From Hive(s):	Product Harvested: Date: Amount: From Hive(s):
Product Harvested: Date: Amount: From Hive(s):	Product Harvested: Date: Amount: From Hive(s):
Product Harvested: Date: Amount: From Hive(s):	Product Harvested: Date: Amount: From Hive(s):
Product Harvested: Date: Amount: From Hive(s):	Product Harvested: Date: Amount: From Hive(s):
Product Harvested: Date: Amount: From Hive(s):	Product Harvested: Date: Amount: From Hive(s):
Product Harvested: Date: Amount: From Hive(s):	Product Harvested: Date: Amount: From Hive(s):

Product Harvested: Date: Amount: From Hive(s):	Product Harvested: Date: Amount: From Hive(s):
Product Harvested: Date: Amount: From Hive(s):	Product Harvested: Date: Amount: From Hive(s):
Product Harvested: Date: Amount: From Hive(s):	Product Harvested: Date: Amount: From Hive(s):
Product Harvested: Date: Amount: From Hive(s):	Product Harvested: Date: Amount: From Hive(s):
Product Harvested: Date: Amount: From Hive(s):	Product Harvested: Date: Amount: From Hive(s):
Product Harvested: Date: Amount: From Hive(s):	Product Harvested: Date: Amount: From Hive(s):
Product Harvested: Date: Amount: From Hive(s):	Product Harvested: Date: Amount: From Hive(s):

BEE BUSINESS INCOME

Here is where you keep track of your bee related income! Whatever you sell that comes from the bees, keep track of it and weigh your income against your expenses to see if you made a profit! Remember, I would grow sustainably and slowly. Don't get too big to fast! You may not be interested in making money from your bees, but most of us are, if anything so you can buy more bee stuff! I have included a column for the name and number of the buyers, this is a great way to build your bee network!

Product Sold	Amount and Price	Name and Number of Buyer

Product Sold	Amount and Price	Name and Number of Buyer

Product Sold	Amount and Price	Name and Number of Buyer

Product Sold	Amount and Price	Name and Number of Buyer

Product Sold	Amount and Price	Name and Number of Buyer

Product Sold	Amount and Price	Name and Number of Buyer

SWARM TRAP LOG

Earlier in the logbook I mentioned that catching a swarm is not a great way to start beekeeping. The reason is simple, if you don't have success catching a swarm, you won't get to use all that pricy equipment you just bought, and never have the chance to start learning the lessons of beekeeping. However, swarm trapping is a great way to continue your journey of beekeeping by catching free "wild" bees. Let's face it, buying bees is expensive, and wild bees may have better genetics adapted to your local environment. Or heck, they may just be a neighbor's bees that swarmed and were bought from the same place you bought your nuc or package!

There are lots of swarm trap designs on the internet to choose from. You can buy a swarm trap or build your own! There are several great online forums on Facebook and elsewhere to get into swarm trapping. You will probably start by having a "safety" swarm trap near your own bee yard to potentially catch your own bees if they swarm without you knowing. If you have a large enough property, you may place swarms in key locations to catch more bees.

Sooner or later, you may get the swarm trap fever and start asking family and friends, even complete strangers if you can place a swarm trap on their property! This log is to help you keep track of all the swarm trap chaos! Make it organized, and you won't regret it! Usually if you are setting up a swarm trap in a new area, it is common courtesy to work a deal with the property owner. Maybe you offer them a free jar of honey, or the promise of honey if a swarm is actually caught! Being a good beekeeper goes a long way in your community! Don't give us a bad name!

Swarm Trap Type: Contains: Lure/Bait Used: Property Owner: Contact Info: General Location: Date Placed on Property: Dates Checked: Removed from Property on:	Swarm Trap Type: Contains: Lure/Bait Used: Property Owner: Contact Info: General Location: Date Placed on Property: Dates Checked: Removed from Property on:
Swarm Trap Type: Contains: Lure/Bait Used: Property Owner: Contact Info: General Location: Date Placed on Property: Dates Checked: Removed from Property on:	Swarm Trap Type: Contains: Lure/Bait Used: Property Owner: Contact Info: General Location: Date Placed on Property: Dates Checked: Removed from Property on:

Swarm Trap Type: Contains: Lure/Bait Used: Property Owner: Contact Info: General Location: Date Placed on Property: Dates Checked: Removed from Property on:	Swarm Trap Type: Contains: Lure/Bait Used: Property Owner: Contact Info: General Location: Date Placed on Property: Dates Checked: Removed from Property on:
Swarm Trap Type: Contains: Lure/Bait Used: Property Owner: Contact Info: General Location: Date Placed on Property: Dates Checked: Removed from Property on:	Swarm Trap Type: Contains: Lure/Bait Used: Property Owner: Contact Info: General Location: Date Placed on Property: Dates Checked: Removed from Property on:

Swarm Trap Type: Contains: Lure/Bait Used: Property Owner: Contact Info: General Location: Date Placed on Property: Dates Checked: Removed from Property on:	Swarm Trap Type: Contains: Lure/Bait Used: Property Owner: Contact Info: General Location: Date Placed on Property: Dates Checked: Removed from Property on:
Swarm Trap Type: Contains: Lure/Bait Used: Property Owner: Contact Info: General Location: Date Placed on Property: Dates Checked: Removed from Property on:	Swarm Trap Type: Contains: Lure/Bait Used: Property Owner: Contact Info: General Location: Date Placed on Property: Dates Checked: Removed from Property on:

Swarm Trap Type: Contains: Lure/Bait Used: Property Owner: Contact Info: General Location: Date Placed on Property: Dates Checked: Removed from Property on:	Swarm Trap Type: Contains: Lure/Bait Used: Property Owner: Contact Info: General Location: Date Placed on Property: Dates Checked: Removed from Property on:
Swarm Trap Type: Contains: Lure/Bait Used: Property Owner: Contact Info: General Location: Date Placed on Property: Dates Checked: Removed from Property on:	Swarm Trap Type: Contains: Lure/Bait Used: Property Owner: Contact Info: General Location: Date Placed on Property: Dates Checked: Removed from Property on:

Swarm Trap Type: Contains: Lure/Bait Used: Property Owner: Contact Info: General Location: Date Placed on Property: Dates Checked: Removed from Property on:	Swarm Trap Type: Contains: Lure/Bait Used: Property Owner: Contact Info: General Location: Date Placed on Property: Dates Checked: Removed from Property on:
Swarm Trap Type: Contains: Lure/Bait Used: Property Owner: Contact Info: General Location: Date Placed on Property: Dates Checked: Removed from Property on:	Swarm Trap Type: Contains: Lure/Bait Used: Property Owner: Contact Info: General Location: Date Placed on Property: Dates Checked: Removed from Property on:

Swarm Trap Type: Contains: Lure/Bait Used: Property Owner: Contact Info: General Location: Date Placed on Property: Dates Checked: Removed from Property on:	Swarm Trap Type: Contains: Lure/Bait Used: Property Owner: Contact Info: General Location: Date Placed on Property: Dates Checked: Removed from Property on:
Swarm Trap Type: Contains: Lure/Bait Used: Property Owner: Contact Info: General Location: Date Placed on Property: Dates Checked: Removed from Property on:	Swarm Trap Type: Contains: Lure/Bait Used: Property Owner: Contact Info: General Location: Date Placed on Property: Dates Checked: Removed from Property on:

Swarm Trap Type: Contains: Lure/Bait Used: Property Owner: Contact Info: General Location: Date Placed on Property: Dates Checked: Removed from Property on:	Swarm Trap Type: Contains: Lure/Bait Used: Property Owner: Contact Info: General Location: Date Placed on Property: Dates Checked: Removed from Property on:
Swarm Trap Type: Contains: Lure/Bait Used: Property Owner: Contact Info: General Location: Date Placed on Property: Dates Checked: Removed from Property on:	Swarm Trap Type: Contains: Lure/Bait Used: Property Owner: Contact Info: General Location: Date Placed on Property: Dates Checked: Removed from Property on:

Swarm Trap Type: Contains: Lure/Bait Used: Property Owner: Contact Info: General Location: Date Placed on Property: Dates Checked: Removed from Property on:	Swarm Trap Type: Contains: Lure/Bait Used: Property Owner: Contact Info: General Location: Date Placed on Property: Dates Checked: Removed from Property on:
Swarm Trap Type: Contains: Lure/Bait Used: Property Owner: Contact Info: General Location: Date Placed on Property: Dates Checked: Removed from Property on:	Swarm Trap Type: Contains: Lure/Bait Used: Property Owner: Contact Info: General Location: Date Placed on Property: Dates Checked: Removed from Property on:

Swarm Trap Type: Contains: Lure/Bait Used: Property Owner: Contact Info: General Location: Date Placed on Property: Dates Checked: Removed from Property on:	Swarm Trap Type: Contains: Lure/Bait Used: Property Owner: Contact Info: General Location: Date Placed on Property: Dates Checked: Removed from Property on:
Swarm Trap Type: Contains: Lure/Bait Used: Property Owner: Contact Info: General Location: Date Placed on Property: Dates Checked: Removed from Property on:	Swarm Trap Type: Contains: Lure/Bait Used: Property Owner: Contact Info: General Location: Date Placed on Property: Dates Checked: Removed from Property on:

Swarm Trap Type: Contains: Lure/Bait Used: Property Owner: Contact Info: General Location: Date Placed on Property: Dates Checked: Removed from Property on:	Swarm Trap Type: Contains: Lure/Bait Used: Property Owner: Contact Info: General Location: Date Placed on Property: Dates Checked: Removed from Property on:
Swarm Trap Type: Contains: Lure/Bait Used: Property Owner: Contact Info: General Location: Date Placed on Property: Dates Checked: Removed from Property on:	Swarm Trap Type: Contains: Lure/Bait Used: Property Owner: Contact Info: General Location: Date Placed on Property: Dates Checked: Removed from Property on:

SWARM TRAP SUPPLIES LIST

Item Purchased	Price	Purchased From	Date Purchased

QUEEN REARING

As your beekeeping progresses, you may begin to get into queen rearing, either for your own apiary for starting nucs, splitting hives, saving a queenless colony, or for adding a queen to a "wild" swarm when the queen wasn't captured. You may even begin to raise queens for other local beekeepers. Use this section to record your research, the method used, queens started, and successful queens made:

Queen Rearing Methods: There are several methods used in queen rearing. As a beekeeper, it is up to you to research and learn and determine which methods work best for you. Once you find the method you would like to try, record the steps below:

Queen Rearing Method 1 Research:

Queen Rearing Method 2 Research:

Queen Rearing Method 3 Research:

Queen Rearing Method 4 Research:

QUEEN REARING SUPPLIES LIST

Item Purchased	Price	Purchased From	Date Purchased

Queen Rearing Method: **Date Started:** **Queen Cells Started:** **Successful Unmated Queens:** **Date:** **Further Actions/Notes:**	**Queen Rearing Method:** **Date Started:** **Queen Cells Started:** **Successful Unmated Queens:** **Date:** **Further Actions/Notes**
Queen Rearing Method: **Date Started:** **Queen Cells Started:** **Successful Unmated Queens:** **Date:** **Further Actions/Notes**	**Queen Rearing Method:** **Date Started:** **Queen Cells Started:** **Successful Unmated Queens:** **Date:** **Further Actions/Notes**

Queen Rearing Method: **Date Started:** **Queen Cells Started:** **Successful Unmated Queens:** **Date:** **Further Actions/Notes:**	**Queen Rearing Method:** **Date Started:** **Queen Cells Started:** **Successful Unmated Queens:** **Date:** **Further Actions/Notes**
Queen Rearing Method: **Date Started:** **Queen Cells Started:** **Successful Unmated Queens:** **Date:** **Further Actions/Notes**	**Queen Rearing Method:** **Date Started:** **Queen Cells Started:** **Successful Unmated Queens:** **Date:** **Further Actions/Notes**

Queen Rearing Method: **Date Started:** **Queen Cells Started:** **Successful Unmated Queens:** **Date:** **Further Actions/Notes:**	**Queen Rearing Method:** **Date Started:** **Queen Cells Started:** **Successful Unmated Queens:** **Date:** **Further Actions/Notes**
Queen Rearing Method: **Date Started:** **Queen Cells Started:** **Successful Unmated Queens:** **Date:** **Further Actions/Notes**	**Queen Rearing Method:** **Date Started:** **Queen Cells Started:** **Successful Unmated Queens:** **Date:** **Further Actions/Notes**

Queen Rearing Method: **Date Started:** **Queen Cells Started:** **Successful Unmated Queens:** **Date:** **Further Actions/Notes:**	**Queen Rearing Method:** **Date Started:** **Queen Cells Started:** **Successful Unmated Queens:** **Date:** **Further Actions/Notes**
Queen Rearing Method: **Date Started:** **Queen Cells Started:** **Successful Unmated Queens:** **Date:** **Further Actions/Notes**	**Queen Rearing Method:** **Date Started:** **Queen Cells Started:** **Successful Unmated Queens:** **Date:** **Further Actions/Notes**

Queen Rearing Method: **Date Started:** **Queen Cells Started:** **Successful Unmated Queens:** **Date:** **Further Actions/Notes:**	**Queen Rearing Method:** **Date Started:** **Queen Cells Started:** **Successful Unmated Queens:** **Date:** **Further Actions/Notes**
Queen Rearing Method: **Date Started:** **Queen Cells Started:** **Successful Unmated Queens:** **Date:** **Further Actions/Notes**	**Queen Rearing Method:** **Date Started:** **Queen Cells Started:** **Successful Unmated Queens:** **Date:** **Further Actions/Notes**

BEE FEEDING RECIPES

This section is to help you get started making basic bee supplements, mainly sugar based supplements to help your bees survive through times of dearth, the winter, and to help young hives draw comb and build up more brood. Remember, these are supplements, and not meant to be the main source of food for your bees. Always leave enough honey and pollen stores for your bees, and if you have extra frames of honey, you can store them and feed back to your bees when needed. Never give sugar feed to your bees when you're going to harvest the honey! Most beekeepers feed their bees at some point during the year, but there are some who will not. I look at my bees as livestock, even though cows and goats are eating grass, you still need to provide supplements when needed! Never feed honey to your bees that was not made in your bee yard by your bees! Feeding store bought honey or even honey from another local beekeeper is bad beekeeping! You could be needlessly spreading disease and bee viruses. These are the "basic recipes", some keepers like to add a few drops of essential oils and Honey-B-Healthy to any recipe mentioned below. You decide what works best for you and your bees!

1:1 Sugar Syrup:

This is a one to one ratio of sugar syrup to water. You can use 1:1 sugar syrup in the early spring to help encourage bees to build and draw out comb and stimulate brood rearing. It's also great to use in times of a dearth, or early fall, after you are done harvesting any honey meant for humans. There is some debate in whether you should measure by weight or volume, but they are so close, it is essentially the same! For small scale, I usually buy a standard 4 lbs. bag or cane sugar and mix with 4 pounds of water, or to make it easy a half gallon of water. Technically a gallon of water is 8.3 pounds, so if you want to add another cup of water to your half gallon, go for it!

After stirring well, this makes about a gallon of sugar syrup, perfect amount for 1 hive. I prefer to use plastic top feeders when I feed, but others do open feeding.

2:1 Sugar Syrup

You guessed it! 2:1 sugar syrup is 2 parts sugar to 1 part water. This syrup is best used in the early fall, to encourage food storage. This just a supplement, you should still leave your bees plenty of honey! The 2:1 ratio is easier for the bees to evaporate the water and store the sugar in the colder season. You will need 8 pounds of sugar for every half gallon! (add 2 more cups if you want). A half gallon of water will not allow 8 pounds of sugar to dissolve into it at room temperature. You need to heat the water up to just below boiling and slowly add in the sugar until it dissolves. Heating up the water expands the spaces between the water molecules and allows more sugar to dissolve in this extra space. When the water cools back down, you have created a supersaturated solution, and the "extra" sugar will remain in solution for a long period of time.

Bee Candy:

Once the temperature falls below 50 degrees F, bees prefer bee candy over 2:1 syrup. It just takes to much valuable energy to lower the moisture content when its cold. The bees need that energy to stay warm. There are lots of ways to make bee candy, some require cooking, but I prefer the no cook method. Its simple! Just mix 1 oz of water for every pound of sugar. Mix thoroughly, then place in a cooking pan lined with parchment paper. Level out the sugar mixture and let dry overnight. That's it! Place bee candy above your hive, using a spacer. Once the temperature climbs back above 50, in the early spring, you can switch to 1-1 sugar syrup if desired. Remember, this is a backup, bees really need an entire super at least of honey to overwinter!

Mountain Camp Method

This is about the easiest way to feed bees in the winter, but I personally prefer to use bee candy. The mountain camp method is literally just pouring sugar straight from the bag onto newspaper above your super and using a spacer. You can also flip your inner cover and pour sugar onto the cover if needed. The main benefit of this method is its easy! Say you don't have time to make bee candy, but notice your bees are out of most of their honey, use the mountain camp method! It's better to feed than let your bees starve. The sugar also will absorb moisture in the hive to some extent, which creates a better winter hive environment. If you have an unusually warm winter in your area, which is becoming more and more common, your bees will be more active, which means they will use more food! If your bees die in late February or early March, it is usually due to moisture problems or starvation. There may be pollen during these months, but not much nectar for most areas.

BEE FEEDING RECIPES:

(This section is to write your own bee feeding recipes)

BEE PRODUCT RECIPES

Here are some great and simple recipes to start using to make value added products from your bees. I have personally made all these products and used, given away to friends and family, and sold at local markets and fairs. I have also included some blank recipe pages, so you can add future recipes you find! Happy making!

Bottled Honey:

Not so much a recipe, but here are a few tips! When starting out, or if you only have a small amount of honey to give away or sell, start with small jars and bottles. Even 2 oz and 4 oz jars sell! Try to have a wide variety of sizes to fit every budget! Don't be afraid to approach local businesses to see if they are willing to sell your honey! I sold my honey in a store my very first year! Harvesting at different times of the season means you will most likely have different honey varieties! Keep track of the blooms in your area to get a good idea of your honey variety!

Cut Comb Honey:

Again, not really a "recipe", the bees are doing all the work here, you just have to give them the right stuff to do it and then package it in the end! Cut comb is 100 percent pure honey and comb, the wax is digestible, and people love buying it! It's a sure sign that people are getting local honey, you can't fake cut comb! To help your bees make cut comb, you just need to insert wax foundation without any wire supports and then give your bees time to draw the comb and, harvest nectar, and make honey! When the honey is capped, it is harvest time. It is best to let your bees make cut comb early in the season, when they are ready to draw comb! There are special cutting squares that make perfect size combs for packaging, but you can also just use a knife. Cut comb goes great on a cracker, bread, biscuit, or just as a stand alone treat. Even with a first year hive, you can usually harvest 1 or 2 frames of cut comb the first year without affecting the overall success of the colony. Just don't be too greedy the first year. The downside is you are not able to reuse the drawn comb the bees have built for next year.

Creamed Honey:

Creamed honey is honey that has been processed to control the crystallization process of honey. Normally when honey crystalizes, it forms large crystals that are jagged and pointy, and not appetizing on the taste buds! Creamed honey is made with a starter batch of creamed honey, that is essentially crystallized honey that is only small and delicate crystals. These small crystals encourage other similar crystals to form. The result is honey with a very smooth texture, somewhere between cream cheese and butter. It tastes just like honey (because that's all that it is) but is much more spreadable. My wife loves creamed honey on biscuits, and when we are down to our last jar, we cherish it! To make this yummy treat, order a starter batch of creamed honey online, and mix your liquid honey with your starter at about a 10 to 1 ratio. Next mix well and then store your creamed honey batch in a cool place, around 50 degrees F. A shed or barn in late fall, or a cool basement work well. You may have to wait 2-3 weeks for your honey to become creamed honey, but as long as you save a few jars, you have creamed honey starter for a future batch! You can also make your own starter if you have crystalized honey. Just put the crystalized honey in a blender or food processor and then grind the crystals as small as possible with a mortar and pestle. Then add your homemade starter to liquid honey to start the process!

Lip Balm:

One of my favorite products to make! This is a great seller at markets all times of the year, and an easy and simple family project. Makes a great gift as well, who doesn't need lip balm?

Ingredients:

- 2 tablespoons of beeswax (melts faster if shredded, or in small pieces)
- 2 tablespoons of coconut oil
- 2 tablespoons of shea butter
- Essential oil – pick your favorite "flavor". Peppermint, tangerine, and lemon are all big hits!

Mix beeswax, coconut oil, and shea butter and melt in a double boiler. When melted, remove from heat and add in essential oil. While still melted, use a pipette dropper to fill up lip balm tubes and let cool for several hours.

Hand and Foot Moisturizer Bar:

Ingredients: 2 oz beeswax, 2 oz cocoa butter, 4 oz shea butter, and essential oil. Go ahead and choose what soothing essential oil you would like to use! First you double boil the beeswax with the cocoa butter. Once both are melted and clear, add in the shea butter, and stir. Once everything is melted again, add in your essential oil, but not until the heat is off. Adding essential oil to a hot mixture will cause it to vaporize, and it won't make it into your product! Pour melted mixture into your molds and let sit 24 hours.

100% Beeswax Candles

Making 100 percent beeswax candles is simple. All you need is beeswax, molds, and candle wicks. Research which candle wick size is needed, based on the diameter of the candles you are making. Melt beeswax until liquid, and then dip your wick into the melted wax. Pour beeswax into prepared mold, with wick preplaced, and then pour in beeswax and let cool. Best to let cool overnight. Candle molds can be expensive, but cupcake pans can work for simple candles! Spraying olive oil lightly, or mold release, can help candle removal immensely!

Beeswax and Coconut Oil Candles

100 percent beeswax candles are beautiful, but they take a lot of beeswax! They burn decently, but if you want to make the most out of a limited supply of beeswax and want a slightly slower burn rate for your candles, then just mix beeswax and coconut oil in a 50/50 ratio and melt on your double boiler system. Dip your wicks in the mixture and then pour your melted wax and oil into containers. Make sure to add the wick BEFORE you pour. Mason jars work well as containers and make great gifts! It is difficult to use a beeswax and coconut oil mixture in candle molds; this recipe is best used for container-based candles.

Holiday Ornaments

Holiday ornaments are fun and simple! Just melt beeswax in a double boiler and then pour into holiday themed molds. Cookie molds and the precious "brown bag" molds work great. After pouring half of the mold depth, add your ornament hangers for support. This can be ribbon, twine, yarn, or metal. Then finish pouring and let cool overnight!

RECIPES

FIND AND MAKE YOUR OWN BEE BASED RECIPES AND RECORD HERE:

RECIPES

RECIPES

RECIPES

GENERAL NOTES

GENERAL NOTES

www.ingramcontent.com/pod-product-compliance
Lightning Source LLC
Chambersburg PA
CBHW081002170526
45158CB00010B/2874